From Ice to Rain

From Ice to Rain

Marlene Reidel

 Carolrhoda Books, Inc., Minneapolis

LIBRARY OF CONGRESS CATALOGING IN PUBLICATION DATA

Reidel, Marlene.
  From ice to rain.

    (A Carolrhoda start to finish book)
    Ed. for 1974 published under title: Vom Eis zum Regen.
    SUMMARY: Describes the cycle in which ice on the
  pond melts into water, which in turn evaporates into
  water vapor, which collects into clouds, which produces
  rain and snow.

    1. Hydrologic cycle—Juvenile literature. [1. Hydro-
  logic cycle. 2. Water. 3. Precipitation (Meteorology)]
  I. Title.

  GB848.R4413 1981   551.48   81-19
  ISBN 0-87614-157-2

        2   3   4   5   6   7   8   9   10   86   85   84   83   82

From Ice to Rain

It is winter.

A thick, solid layer of ice covers the lake.

The ice is thick enough to hold up the children
as they skate across it.

But underneath the ice there is still water.

When spring comes,
the ice begins to melt in the warm weather.
It becomes water again, but not all at once.
Blocks of ice float on the surface of the water
along with the ducks.
Ice floats because it is lighter than water.

Now the summer sun shines brightly.

There is no ice left in the water.

The children sail their paper boats in a puddle.

But each day the puddle gets smaller.

Soon there is no water left in the puddle.

Where did it go?

The sun has heated the water.

As it got warmer, tiny particles of water called **water vapor** rose into the air.

We say the water has **evaporated**.

The water vapor can't be seen, but it is all around us in the air.

As the water vapor rises, the air becomes cooler.
The cool air makes the water vapor gather
into larger particles called **water dust**.
The water dust then gathers together to form clouds.

When the air becomes cool enough,
the water dust turns into water drops.
The air can no longer hold so much water,
and the drops fall to the ground.
It is raining.

In the cool fall we often see mist rising
from lakes and ponds.
During the day
water vapor rises into the warm air.
But at night the air becomes cool.
The vapor turns into water dust
before it has a chance to rise.
In the morning we see the water dust rising.
It looks like a cloud near the ground.

Soon it is wintertime and the air is very cold.

The water in the rain barrel has frozen into ice.

Now the water dust in the air

doesn't fall in the form of rain.

Because the air is so cold,

the water turns into tiny frozen crystals and it snows!

**Marlene
Reidel**

MARLENE REIDEL was born in lower Bavaria and was raised on an isolated farm called *Krottenthal*. She is the oldest of seven children.

Ms. Reidel studied ceramics as a girl and then went on to attend the Academy of Fine Arts in Munich. She has written and illustrated many children's books and has received numerous honors and awards for her work, including the German Youth Book Prize, the Most Beautiful German Book of the Year award, the Culture Prize of Eastern Bavaria, and the Special Prize of the German Academy for Children's and Youth Literature.

THE CAROLRHODA

START

From Beet to Sugar

From Blossom to Honey

From Cacao Bean to Chocolate

From Cement to Bridge

From Clay to Bricks

From Cotton to Pants

From Cow to Shoe

From Dinosaurs to Fossils

From Egg to Bird

From Egg to Butterfly

From Fruit to Jam

From Grain to Bread

From Grass to Butter

From Ice to Rain

From Milk to Ice Cream

From Oil to Gasoline

From Ore to Spoon

From Sand to Glass

From Seed to Pear

From Sheep to Scarf

From Tree to Table

TO FINISH

BOOKS